CHRISTMAS HOUSES

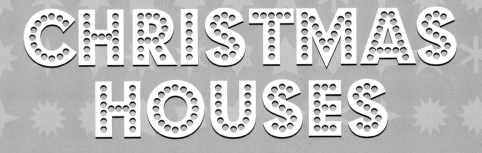

CHRISTMAS HOUSES

Rob Harris, John Hartley & Sam Manning

BANTAM PRESS

LONDON · NEW YORK · TORONTO · SYDNEY · AUCKLAND

TRANSWORLD PUBLISHERS
61–63 Uxbridge Road, London W5 5SA
A Random House Group Company
www.rbooks.co.uk

First published in Great Britain in 2007 by Bantam Press
an imprint of Transworld Publishers

A CIP catalogue record for this book is available from the British Library.

ISBN 9780593060322

Addresses for Random House Group Ltd companies outside the UK
can be found at: www.randomhouse.co.uk
The Random House Group Ltd Reg. No. 954009

The Random House Group Limited supports The Forest Stewardship
Council (FSC), the leading international forest certification organization. All our
titles that are printed on Greenpeace-approved, FSC-certified paper carry the FSC logo.
Our paper procurement policy can be found at:
www.rbooks.co.uk/environment

Text design: www.carrstudio.co.uk
Printed and bound in Great Britain by Butler and Tanner, Frome

2 4 6 8 10 9 7 5 3 1

CONTENTS

INTRODUCTION

It's been dark since 3.30 p.m., it's cold and it's wet, yet more than fifty people line the pavements of a tiny cul-de-sac. Drawn like sunflowers to the light, they are here to bask in the warmth of the Christmas House. This is extreme decorating, a glorious and sometimes perilous sport, but one that lights up the lives of many a household (and their neighbours').

The more conservative Brit may baulk at this gloriously tacky form of exhibitionism, but house-bling is enjoying a growing popularity. It is a Yuletide movement with its star of wonder decidedly in the ascendant, blazing a blur of coloured lights across our horizons at the darkest time of the year.

This book is a celebration of the sheer wonder of external Christmas decoration. It collects together some of the best Christmas Houses to be found across the UK and gives a glimpse into the lives of the people behind these creations. Why do they do it? How do they do it? How do their neighbours feel about it? (And did that snowman just wink at me?)

7

Chapter 1

STORIES OF CHRISTMAS

Who are the people behind the crazy façades? Here we meet some of the unacknowledged artists who are pioneering this method of jaw-dropping personal expression.

Huntington, Staffordshire

The Glitter-cage House

Genre: End of terrace
Style: Disco, disco!

This tip-top, top-drawer Christmas House takes an unusual approach. It features a delightfully crazy light display with an original glitter-cage entrance. Creators Mr and Mrs Y begin setting up in late November and are clearly very practised at achieving such a detailed display in so short a time. The trick seems to be military organization and the family work together like a well-oiled, bling-bringing, Homer-inflating machine: Mr Y, who is disabled, tackles indoors whilst Mrs Y marshals their two sons to work outdoors. The family have been exhibiting for at least ten years and in that time they have fully embraced developments in lighting technology, moving from old-fashioned, multi-coloured tungsten bulbs to modern-day rope lighting, with great effect.

Creativity 7 Lightometer 7

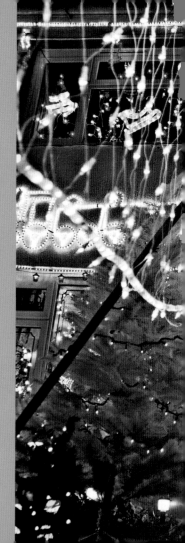

'It's silly'

— *Kids by the fence.*

The house forms a beacon on the A34 and, as with another Christmas story, many hundreds of visitors are guided from afar by the lights on the horizon to look and take photographs (or something like that!). Local kids congregate in the flattering glow of the fence on frosty December nights.

What you can't see:

The dazzling display flashes once a second and the auditory sense is catered for with outside speakers playing Christmas pop (or 'Criss-pop') classics on a loop.

13

Binstead, Isle of Wight

Genre: Detached house
Style: Uniquely handcrafted

Mr B of Binstead wanted to light up his house ever since he watched the movie *National Lampoon's Christmas Vacation*. He finished building this house himself about eight years ago – and then clearly had a bit of time on his hands, so he and his wife set about decorating it. The initial display was a simple affair (by comparison) and seemed a fitting celebration of having created one's own home. However, since then the involvement in the house decorations themselves has grown . . . and grown. No trips to the DIY shop to buy ready-made figurines for Mr B – 'I just don't want to do that!' – this entire display is bespoke and home-made.

At the beginning of the Christmas season, an opening ceremony takes place, attended by a crowd of admirers. Featured celebrities have included Geoff Hughes from TV's *Keeping up Appearances*, Mr Blobby and Santa, and the switch-on is accompanied by carols played stoically by a local children's brass band. Mr and Mrs B have been declared Detached-House Winner by *Wight In Lights* for three years running and have raised around £10,000 for Marie Curie Cancer Care.

The Workshop and below a detail showing Lenny and Herbert hard at work

Once the decorations are in place for the current Christmas season, Mr B begins planning for the following year. Realization of the creative vision takes most of the year, as every figure is designed and built individually, inventively mobilized by windscreen-wiper motors. Come September, Mr B can be seen clambering over his roof on Saturdays, getting the first elements in position. Popular snowman band The Snowies are a regular feature, providing the soundtrack for the season and over Christmas 2006, thanks to diminutive volunteers, Santa's workshop was open day and night making and wrapping presents and tying bows.

the Snowies

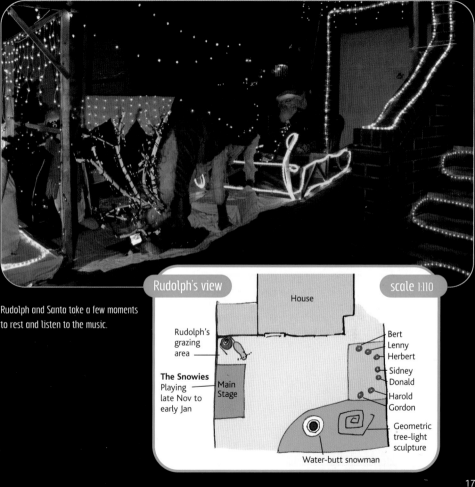

Rudolph and Santa take a few moments
to rest and listen to the music.

Rudolph's view

scale 1:110

House

Rudolph's
grazing
area

The Snowies
Playing
late Nov to
early Jan

Main
Stage

Bert
Lenny
Herbert

Sidney
Donald

Harold
Gordon

Geometric
tree-light
sculpture

Water-butt snowman

Chelmsford, Essex

The Rock 'n' Roll House

Genre: Mid-terrace
Style: Definitely not rockabilly!

Rockin' Around the Christmas Tree

ROCK 'N' ROLL

Seven o'clock rock
– an artist's impression.

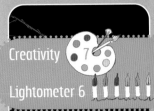
This remarkable 'Rocking Around The Christmas Tree' rock 'n' roll special was created in an impressively short preparation time of two days! Resident Mr J runs his own scaffolding business and uses his expertise in this field to mount the display so efficiently. He finished re-cladding the front of the house just before these photos were taken and consequently only managed to complete the decorations by 4 December (a *little* later than usual).

Mr and Mrs J run a busy rock 'n' roll night in Essex and have found their enthusiasm spilling over into their lighting displays, which they base around a year-to-year rotating theme of their favourite rock 'n' roll hits. This year it's the Brenda Lee classic 'Rockin' Around The Christmas Tree'; last year was Elvis's 'Blue Christmas'. At seven o'clock every evening, the snow machine kicks in to the sound of this year's theme song and the street comes alive! There is hardly any decoration inside the house, as all energy is focused on the exterior. Mr J creates all the light bulb lettering himself and aspires to create personalized figurines and components in future years.

Mr and Mrs J have been exhibiting for five years and collect donations for the Essex Air Ambulance. The couple's future plans could involve their own rock 'n' roll pub in Chelmsford, which would of course provide much wider scope and space for even more rocking Christmas lighting.

Great Baddow, Essex

The Spotty House

The jaunty decorations on this house also help to raise money for Essex Air Ambulance, for whom resident Mr P is a volunteer speaker. We particularly enjoy the pencil drawing style here; dotty pointillism is combined with linear definition to great effect.

Mr P's highlight of the last season was when celebrity Barry Daines (ex-Tottenham Hotspur goalkeeper) kicked off the lighting-up ceremony to a roar from the crowd.

Genre: Semi-detached
Style: Daft and dotty

Bromley, South London

Genre: Detached house
Style: Festive jungle

This fantastic house in Bromley gets buried under its own Santa mountain every December. The decorations include streaming lights, ladders, icicles and good old end-of-the-pier-style strings of coloured bulbs. The glowing jungle in front of the house has many surprises peeping out from behind the leaves. One visitor claims to have seen an LED caveman, but they might have been making it up...

We spotted an astonishing sixty-five plus Santas (plural 'Santi'?) featured with trains, sleighs, unicycles, reindeer, a ladder and even standing in a wooden box.

Not a bad haul.

Santometer

Roof: 5

Front: 20

Garden: 40+

23

Weston-super-Mare, Somerset

Genre: Semi detached house
Style: Noël of the naked noggin

Featured on this spectacular plot there are not one, not two, but three Christmas Houses. Look closely and you will spot a Christmas Wendy House in the garden and an unusual grotto, converted from a plant stall, standing by the gate (*see detail*). We were also very impressed by a rare sighting of snowmen on stilts.

This decorator has put his heart, soul and body hair into his decorations and fund-raising activities. He grew his beard for a sponsored beard shave, but when it came to trim it off down the pub, one thing led to another and he came back totally bald.

Santometer – 24

The Stories of Christmas

Eltham

Genre: Semi-detached house
Style: Delightful glitzy colour

These decorators started thirteen years ago after a trip to the States and they've been back there for more Christmas lights every year since. (The Santa chimney featured on the opposite page was acquired From Wal-Mart, Orlando.) However, their tip is: if you're interested in getting the best decorations available, then America isn't as promising as it once was. Our British bling is just as good.

Highly Commended
for Outstandingly Tasteful
Christmas Decorations

The Eltham and Blackfen
Illuminations Society

This is a success story all round. Their neighbours love it, especially when Father Christmas arrives for the light-up on 1 December and hands out a small present to all the kids that turn up. To top it off, they were recently awarded a certificate of excellence by the Eltham and Blackfen Illuminations Society.

What you can't see...

Alternating flashing bells clang out Christmas cheer

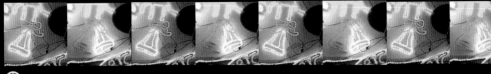

A seasonal tiding is strobed to one and all

Jaunty Santa waves 'peekaboo' from within his inflatable chimney

"It used to be that the kids ran home from school hoping no one saw that they were associated with such a loud house. But, after thirteen years their attitude has changed and now they are helping."

Dorchester, Dorset

Genre: Semi-detached house
Style: Liquorice allsorts

A mesmerizing onslaught of light, shapes and movement brings cheer to this street in Dorchester. Not satisfied with audio-visual impact alone, the owner of this house also indulges her audience by handing out mince pies to visitors.

What you can't see:

The intense concentration needed to separate one shape from another means that a line of viewers stand open-mouthed in awe as they work their way through the visual puzzle that is presented in this front garden.

A visual puzzle

Santometer - Santas 43 , Snowmen 14

The Mad-Eyes House

Genre: Detached house
Style: Wacky winter wonderland

Mr P rotates his charity beneficiary each year. 'I felt inspired to chose Shanklin Town Brass Band this year after seeing an episode of a TV series exploring music theory, in which the pentatonic scale was so very excellently explained. It had been an unfathomable mystery to me for many years.'

Mr P's design philosophy:

1. to include a nativity as a reminder for the reason for Christmas;

2. to have moving lights for children;

3. to create the appropriate background for it all.

Mr P relates how he was surprised by one visitor:

'About three years ago at half eleven at night, I went out to turn the display off and there was someone stood on top of the bank by the road. So I went back indoors for ten minutes to give him time to look at the lights. When I came out again he was stumbling up the path towards the house, so I made my presence known and the conversation went as follows:

Me: Hello, can I help you?

Him: Oh, hello. Is this a pub?

Me: No, it's just a Christmas lights' display.

Him: Oh, I saw all the lights and thought it was a pub. They're very fine lights too.

Me: Thanks.

Him: Oh, by the way, I hope you've no fish in your pond cos I've just fallen in and I would hate to have hurt one.

Me: [*Speechless*]

Him: They're very fine lights, are you collecting for something?

[*He reaches into his pocket and gives me all the soggy loose change he has, and fumbles in his other pocket for his mobile phone.*]

Him: Oh dear, I don't think that will be working. [*Water drips from his mobile.*] Ah well, they're very fine lights but I must be going.

And he staggered off, never to be seen (by me) again.'

33

Denton, Manchester

'We've been putting up decorations for twenty-one years, and we love it.'

Genre: Victorian corner terrace
Style: Exuberant

It all started with the modest idea of putting lights around some Georgian windows – perhaps no more than four or so lights in each pane, just for fun. Emboldened, the next year they added a few lights to the door. Since then, the decorations of this family in Denton have grown steadily every year to become the fantastic Christmas extravaganza featured in these pages. The house is also heavily decorated inside. 'We don't restrict ourselves to the outside!' Mr D decorates enthusiastically all over the house, but he has been allowed to really go to town in the living room (red and gold themed) and the dining room (silver and gold). Some attention is also extended to the downstairs toilet, our photographer reports.

On 1 November there can be up to eighty visitors to the street, celebrating the turn-on, so a conscientious Mr D starts hanging his light display a whole month before – around 1 October. Over the festive period, his warm and bountiful hospitality is such that he invites passers-by to come in, share a coffee and enjoy a tour of the interior.

In part, the display is for two lucky live-in granddaughters and to raise money for The Downs Society. However, it is pretty clear that this family simply love the warmth of Christmas. The Christmas build-up has escalated to involve more cards, more presents (Mrs D is the only one with enough time to think about these) and the buying of decorations throughout the year.

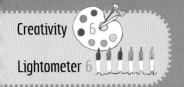

Creativity 6

Lightometer 6

Christmas Day itself begins with midnight mass and reaches a climax with a big family Christmas dinner for twelve.

Then, suddenly, it's all over.

'January feels horrible,' said Mr D.

This big detached house with enclosed driveway, entirely dedicated to lit-up Christmas cheer, can be found in a locality that already boasts a high lighting quotient. Visitors must enter the grounds in order to view the remarkable spectacle and the owners have used this to their advantage, raising many thousands of pounds every year for the Little Havens Children's Hospice from donations. Clearly this is house-bling taken to a professional level, as some of the features announce that they are sponsored by local businesses. Not surprisingly, the attraction draws large numbers of visitors until late into the night.

The house itself appears to be heavily garrisoned by commander-in-chief Santa and a sizeable brigade of *Nutcracker*-style soldiers keeping a watchful eye on the jollity levels. In the forecourt, the display consists of a succession of marquees

...using various festive trompe l'oeils (paintings that fool you into thinking they're real), a traditional nativity, a train montage and a favourite snowman competition (we're not quite sure what happens to the winner – maybe he gets put in the freezer over summer).

An examination of the design style used here must surely note the very high proportion of Lit-Up Moulded Plastic Statuettes. 'They're crackers, those LUMPS.' – quips a passer-by.

Genre: Detached house
Style: Santa freight: a tale of international travel

Many a child on the edge of sleep has wondered how it is that Santa can possibly cover thousands of miles in one night from Melbourne to Rio de Janeiro via Bognor Regis, keeping within the envelope of dark that moves across the globe. Well, the practicalities of such high-speed, magical movement go beyond the scope of this book, but we can see that it's a question that seems to have occupied Mr H a great deal. Here he presents his exploration of Santa-based travel and distribution solutions.

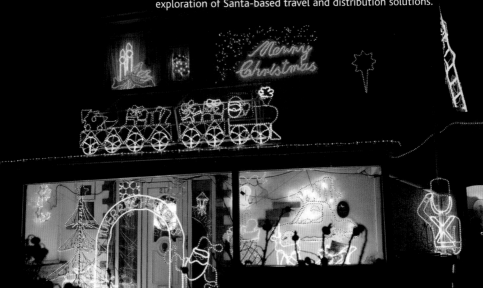

Clearly, Mr H has a good understanding of the industrial scale of gift transportation. The old favourite the steam train features, as well as the nippy run-around motorcar. We can see more trains and carriages around the back of the house, hopefully pulling in to the upstairs window to unload. The final destination of the presents – the tree – completes the distribution showcase.

Brading, Isle of Wight

Genre: Detached house
Style: Surfing a wave of colour

The creation of this shifting wonder is as much a fine art as the lighting display itself. It takes Mr and Mrs W a comfortable couple of weeks in four-hour blocks to construct, thanks to six years of practice and the installation (by an electrician) of large plug boards in set locations. The complications of exhibiting on a sloping site have been overcome by the use of level platforms to support features such as the Crimbo-Pingu-style montage and the half-size model railway.

The lucky grandchildren for whom the lights first began to twinkle clearly love their Christmas treat. Mr W's grandson has been inspired to create his own displays – although the pair don't quite see eye to eye on design ethos. 'He tends to favour eight-foot blow-up cartoon characters like the Grinch and Homer Simpson. I think he's got about eight of them. Yeah, they're a bit over life-size...' says proud Mr W.

Mr W's design philosophy:

 I like the use of colour. Some people like all-white displays, but I don't want to do that. In my displays I like to blend colour and make movement without using extravagant mechanics. I enjoy highlighting the colour that is there, using floodlights rather than flashing bulbs.

 I don't like rope lights.

What you can't see:

A cross-spectrum of colour flows across the fibre-optic lights like small schools of (very slow) fish in the sea, or the wind blowing (slowly) across a grassy plain.

Perhaps with an eye to the future of Christmas houses, the visionary Mr W has taken the unusual route of opting to use almost entirely fibre-optic and LED technology. The benefits of this method of lighting are:

1. Lower electricity consumption, which means a smaller carbon footprint and lower bills.

2. Components are cheaper to buy.

3. Most elements of this display were found in the UK, meaning fewer costly shopping trips to the US.

4. Less bling, more Fabergé. (To decide whether this is a benefit, see Chapter 4.)

Mr W has built up to this display over the last five years, hunting down any fibre-optic pieces that could be used outside – whether designed for this purpose or creatively adaptable. The shift towards this more environmentally friendly approach was motivated partially by the advantages listed above (Mr W now finds that there is no great jump in his electricity bill over the festive season as all his lights run on 12 volts rather than mains voltage) but primarily because he feels it looks better. Simply put, he prefers a subdued display. **Subdued**, *adj.*, toned down; quiet, restrained; dejected, in low spirits; passive. Hmmm.

Mr and Mrs W collect donations for the Isle of Wight children's charity The Kerry Green Trust.

Equivalent daily energy consumption

Coventry, Warwickshire

Genre: Semi-detached house
Style: Chic-tack

This spectacular display can be seen from half a mile away on the darkest night. There's no mistaking those lights in the distance – it's not the glow of the ring road, nor the supermarket car park – it's clearly house-shaped, with eaves and windows picked out in glowing lava-light rope. The Christmas House sits pertly, waiting for Santa's deliveries, with a big asterisk to mark the drop-off point.

The master of design behind this beacon of visual tidiness is rightly proud of his work. Mr B works as an electrical wholesaler, so he's well placed to find out about the best and latest kit. He says that this particular display is relatively small compared to previous years.

This high-altitude illumination can be difficult to set up, so he leaves up much of the trunking all year round. Even so, it took him and three friends to put up the chimney firework (this was a special piece of kit from the supplier to the city council that he invested in as a treat to himself). The result: civic building-scale atop a sleepy family house which is worth every effort.

Creativity 7 Lightometer 4

1. Using Thermographic Imaging Technology Systems

2. Rude Cam: Rumour has it that Rudolph can only see colour in the red spectrum (something to do with that famous nose?).

This house demands more than just one look.

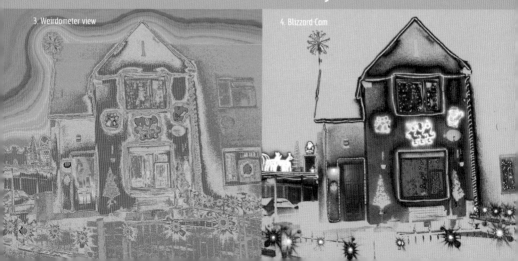

3. Weirdometer view

4. Blizzard-Cam

Chapter 2

IN THE DARK STREETS SHINETH

Blackbird Leys, Oxford

Elephant and Castle, London

Chelsea, London

Keighley, Yorkshire

Wookey Hole, Somerset

Coventry, Warwickshire

Huntington, Staffordshire

Aylesbury, Bucks

Walworth, London

No waiting except buses

Ilkley, Yorkshire

Liverpool, Merseyside

Keighley, Yorkshire

Chapter 3

BIG

Christmas is a time for magical beings to walk the land.

We all know that Santa is assisted by an army of elves and a vast team of reindeer to achieve superhuman feats in just one evening. And now we are beginning to recognize that enormous glowing snowmen are a common feature of our towns and cities in the darker months of the year. Perhaps generations to come will tell traditional Yuletide stories of a seasonal invasion of the giants!

The technical name for these enormities is Fan-Assisted Rigid Tortion Structures or alternatively, Blow-Up Model Santas/snowmen/Simpsons, but these terms are most commonly used in their abbreviated forms . . .

KITSCH?

Classy?

Chapter 4

CLASSY VS KITSCHMAS

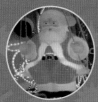

This contemplative and deeply theoretical discussion pitches the two camps of Christmas familiar to any tree-decorating family against each other. Many a glass of sherry has been spilt over this ever-vexing point of philosophy, but here we face the question head on: is it possible to be classy, or should we just embrace the flamboyant gaudiness of the whole thing?

It can take many years to hone one's taste and establish where one's preferences lie. Or, you can cheat.

Simply complete this quiz by choosing A or B as you read the chapter. Count up your answers at the end to determine, once and for all, whether you fall on the side of kitsch or classy.

Classy? or **KITSCH?**

A Westcott, Buckinghamshire

Oxford, Oxfordshire B

'It took a long, long time to create, but I bet you saw this and just went wow!'

Worthing, West Sussex B

'I wouldn't even consider buying a multi-coloured train. To be honest they look tacky.'

Classy? or **KITSCH?**

A Appleford, Oxfordshire

'On a misty night like this it looks really magical.'

A Fulham, London

'Have you seen my present?'

'The lights get turned on 1 December, or earlier if Chelsea are playing at Stamford Bridge, so that all the passing fans can enjoy them'. Other beneficiaries include the owner's twenty-three godchildren.

What you can't see:

An indoor Christmas tree erupts polystyrene snow from its tip – 'The godchildren do tend to make a bit of a mess of it.'

Oxford, Oxfordshire **B**

Classy? or **KITSCH?**

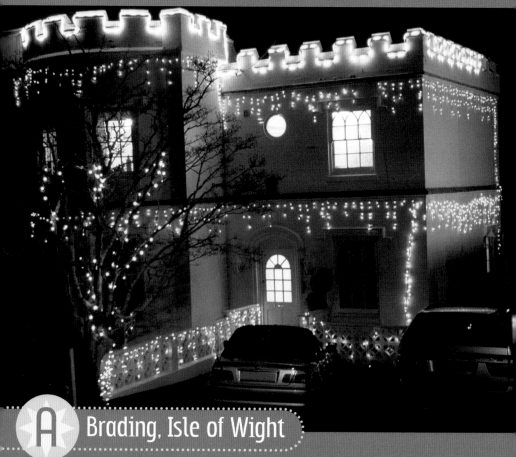

Brading, Isle of Wight

Classy? or **KITSCH?**

Liverpool, Merseyside

Coventry, Warwickshire

Classy? or **KITSCH?**

B Milton Keynes, Buckinghamshire

Only third year decorating

Classy or KITSCH - What's the Verdict?

Mostly Ⓐ

You are absolutely right; it is entirely possible to ooze class from every bulb. After all, where's the artistry in slapping figurines indiscriminately across your façade when you could use beautifully straight lines to enhance the carefully proportioned features of your house? The miracle of Christmas can only truly be expressed through delicately portrayed winter wonderlands (in harmonious colours). And you wouldn't use that tacky approach to decorate your kitchen, would you?

Mostly Ⓑ

Yes, you've got it. The only way to go is to embrace the bling and ditch those white bulbs in favour of every colour of the rainbow. After all, this is Christmas! There's no point being even remotely adult about it, surely child-like glitzy tack is the emblem of the season. How are you supposed to believe in Father Christmas when you're being all grown-up and colour-coordinated? Christmas is a great big tinsel-fest, and you love it.

Mix of Ⓐ and Ⓑ

Torn between the two? Never mind. After all, what are fences for if not to sit on or run rope lights along? Some of the classy houses are magical, but those dancing Santas are amusing too and maybe you enjoy the odd Criss-pop classic. Christmas is a time for the kids and yet those carefully thought-out colour schemes appeal to adults too, so everyone is happy. And of course we must have the classic, cheesy Rudolphs and jolly, musical Frosties along with ground-breaking abstract pointillism. Decision-making is overrated, let's duck out.

'You can see it across the valley from here!'

Chapter 5

NOËL NEIGHBOURHOODS

Sometimes, the most spectacular displays are achieved when a group of close neighbours within a community decide to decorate together.

Long-standing traditions are often formed, with the earliest Noël Neighbourhoods dating back to the sixties.

For some, this is a harmonious arrangement with a street-party atmosphere, while for others a degree of one-upmanship can begin to smoulder. This section is dedicated to the Christmas streets and villages to be found across the country and the dynamics that link the inhabitants.

Turning into this cul-de-sac in Oxfordshire is like suddenly arriving in an astonishing Technicolor Christmas theme park. A tantalizing glimpse of it can be seen from the main road. Blink once and you're on an average village road on a dark winter evening, blink again and there are a thousand rainbow-coloured fireworks in every shape you could think of (although little podgy snowmen and Santas prevail). Flecks of light burst into your vision, softly settling on all surrounding surfaces. The air is full of ho-ho-ho's and the timeless sound of Criss-pop favourites echoes from behind nodding wooden reindeer. Your vision is assaulted by jiggling, flashing lights as you wonder if you're being struck down with a migraine. But no, all is well. This is the glittering spectacle of Long Hanborough.*

* Note: This Noël Neighbourhood holds a special place in our hearts, as it first lit up our interest in Christmas lights and so inspired this book.

House 1

Genre: End of terrace
Style: Mammoth Christmas grotto

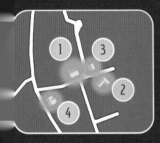

Location plan

This house is definitely the daddy of the neighbourhood. Owners Mr and Mrs S have used their pivotal, end-of-terrace location to maximum effect. Three sides of the house and the entire garden are highly decorated. The majority of their gravelled garden is visible from the road and makes the ideal habitat for any reindeer, tiny Santa, shooting star or fairy (light) that may choose to live there – and many do. Rather than keep the small rear yard as a private area, Mr and Mrs S have thrown open the gate and decorated further.

On the Saturday evening we dropped in, visitors were queuing all round the house for entrance to the yard, the chance to meet Santa in his grotto, and to share a mince pie and cup of mulled wine with the hospitable owners. Apparently, Santa himself had promised to come, but had been caught up in the Christmas rush, so Mr S shyly donned a grey beard and merrily provided an almost indistinguishable stand-in.

In earlier months, the grotto had been a brick shed, but had undergone a Cinderella makeover involving layers of white paint, shining Christmas trees and friendly reindeer-in-residence. Perhaps the summertime residents of the shed were some of the many creatures finding a home in the garden.

All this started as something to entertain children and grandchildren, but it has grown into a spectacular yearly event with participation from the entire community. The display helps collect money for the leukaemia ward at Oxford's John Radcliffe Hospital, where two family members were treated.

As we left, a small amount of (fake) snow drifted down, completing the fantasy.

Genre: Detached bungalow
Style: In memory of George

Dear Christmas Houses

In the last nine years we have spent several thousand pounds on lights, travelling many miles to find ones that nobody else had or had seen, including some we brought back from our holiday in Australia. My husband George died on 1 November last year and it was his enthusiasm that made the lights as good as they were in previous years. He will be remembered for all the hard work he put in to make the garden a lovely place throughout the year (he also won awards in the summer). I will do my best to keep the garden how he would have liked it, but sadly I will not be able to do the lights. So I have sold them locally so that other people will be able to carry on the Christmas spirit.

George wanted Christmas to be an even more special time for the children in the village. He loved to see their smiling faces. But it wasn't only the children who enjoyed them, so did the adults. The idea spread around the village and many other homes were lit up. Hanborough became a very busy place in the lead-up to Christmas, with lots of people coming to see the lights and many others coming to stay with family or friends at the beginning of December just to be there when we had our official switch-on, which was done by local celebrities. A great time was had by all and lots of money was raised for the John Radcliffe Hospital leukaemia ward.

Last year's lights were just a small selection in his memory. The lights featured here were assembled for George by his wife and sons.

House 3

Genre: End of terrace
Style: Outside Santa's workshop gates

House 4

Genre:
End of terrace
Style:
Slightly toned-down
(see p.162)

Melksham, Wiltshire

Genre: Detached house
Style: Of its own!

Melksham has become the local centre for all things Christmas, with houses that fix this quiet market town firmly on the bling-map. The home of two Melksham brothers is one such focal point. These lads have been putting up lights for a number of years and, despite being younger than many we have met, their efforts are easily as good as any more seasoned bling-bringer.

Mr A caught the lighting bug when he was only eleven. His brother also helps, but mainly just with holding the ladders...

His enthusiasm has caught on with his neighbours to such an extent that houses up and down the road illuminate the Christmas nights with displays of lights, decorations, and even an illuminated tractor.

Mr A's interest has also influenced his career choice – he's now an electrician, so was very capable of upgrading the house supply to handle the 90 amps it draws during December.

When he was 20 he undertook the ultimate Christmas lighting pilgrimage – to America, visiting his uncle in Arizona. His 25kg lighting haul from that trip landed him with a meaty luggage surcharge on the plane home! Worth every penny.

Creativity **7**   Lightometer **7**

Other householders add to the impact of this famous display. Together they draw neon-worshippers to this area from far across Wiltshire.

Mrs M also collects decorations from America. She tops off her visitors' experience by offering a sweet to everyone who comes by – or two if you're cheeky!

Mr W decorates his favourite tractor, a Fordson Super Major that he bought from a friend for £100. We are told that the Fordson Super Major is an improved version of the Power Major, produced between 1961 and 1964 in Dagenham.

Melksham is also the home of the notorious, self-styled 'Mr Christmas' for whom every day is Christmas Day, but more on him later!

Littlemore, Oxford

Genre: Modern terrace, cul-de-sac
Style: Sweetie-shop glamour

As you approach this cul-de-sac, you can see the tantalizing, radiant glow of decoration through gaps between houses nearby. Then the dazzle of each display is reflected in the windows of the houses opposite, doubling the impact. It is a quiet residential area and the simple colourful jollity of these creations suggests community warmth and enjoyment of the season. But is that what lies behind the glittering façade? Er, yes! The notion of rivalry is shocking to these decorators: 'Oh no, there's none of that here!'

Mr P's motivation:

'I set up the display every year for my grandchildren and the local community. Some people come from further afield to see the display. I think that it cheers up the area and gets everyone into the Christmas spirit. It makes me feel very proud watching my grandchildren's and other children's faces when they see the lights.'

'I adore Christmas, I do!'
— Mrs M.

Across the road, Mrs M and her daughter Miss C decorate due to a love of Christmas. They began with just one decoration, then bought another, then another... Miss C puts up the decorations in time for 1 December whilst her elderly mother holds the ladder, then they coordinate the switch-on with Mrs P over the road. As Christmas draws nearer, lighting-up hours grow longer, ensuring that Santa will have no problem finding this house on Christmas night, no matter what time he visits.

The light-free neighbours are said to be in favour, but Mrs M regrets that she has been unable to persuade them all to do it, despite her enthusiasm. 'We're all getting on, and some years we feel we can't do it . . . but we love it, so each year we do!'

Wigmore, Kent

Approaching half a century and still going strong!

As the phenomenon of Christmas Houses spreads nationwide, and Noël Neighbourhoods multiply, we take a look back at one the earliest documented Noël Neighbourhoods in Wigmore, Kent.

Credit Graham Smith

The exact date that the decorations were first displayed is unclear. One local resident recalls seeing the lights as a child sometime in the early sixties, when the road was just a dirt track. Exactly which house started the decorations with a wood-and-paint Santa on the roof is lost in the mists of time, but very soon more and more people were joining in with the festivities which reached their peak in the eighties.

Credit Graham Smith

At the height of the lights' fame, a resident electronics expert built an elaborate moving display with Santa sitting on the roof, dropping presents down the chimney. He also constructed a motorized manger (it *rocked,* baby!) for a neighbour. Traffic jams would develop as hundreds of people visited this phenomenon every night. Road closures and the presence of local TV camera crews simply added to the road's notoriety.

Rumours still abound of how the obligation to install a Christmas display was written into the house deeds of new arrivals on the close. One resident had a memorable experience when moving in. He was furious to find what looked like boxes of junk in the attic. He quickly realized that it was a complete display left by the vacating owner. 'When he said he'd leave the lights, I thought he meant a bulb or two in the hallway sockets!'

Nowadays home-made decorations have been overshadowed by industrially produced rope lights and inflatables that have taken up the festive banner. The queues of traffic don't form any more and people who take part in the street's decorations admit that they are no longer at the cutting edge. That these neighbours still put up a display after more than forty years is a remarkable feat of perseverance that just adds to the story of this road.

It's great to see one of the very first ground-breaking community displays still raising an eyebrow with the same low-cost, low-waste decorations. An A for effort, A for creativity, A for longevity!

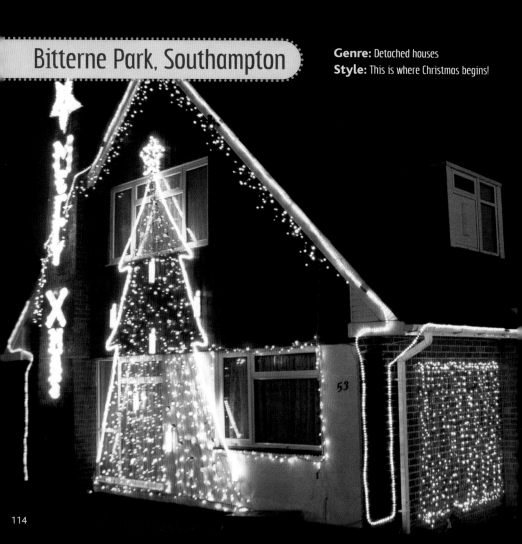

Bitterne Park, Southampton

Genre: Detached houses
Style: This is where Christmas begins!

53

The Christmas light-up started at least fourteen years ago, but these modest beginnings have faded back into the mists of time; nobody can remember how long exactly. Legend has it that it all started with the large tree lit up in Mr N's front garden and one small Christmas tree outside Mr S's house. From these pioneering beginnings, today more than fifty houses in the immediate area participate and the area is widely dubbed 'Tinsel Town' over the Christmas period. Continuous traffic flows by in the lead-up to Christmas and local taxi drivers often take passengers with a little time to spare around 'Tinsel Town' to show them the display. The clear-sighted local council have introduced a one-way system to cope with around 12,000 visitors who come to look at the displays and the cars that convey over 4,000 children to see Santa at a rate of 250 to 300 per night. Apparently, pilgrims travel here from as far abroad as Bournemouth!

Rudolph's view

The epicentre of this high-profile Noël Neighbourhood is a fortuitously located T-junction which forms a natural gathering point.

For his part, Mr N welcomes the local children as Santa, handing out presents donated by everyone in the street from his imposing hand-built grotto. Each child receives a small present and sweets, and tells Father Christmas what they would like to find in their stocking on Christmas morning. In one year alone he managed to give away 3,500 balloons – well done, Mr N! When sweets get low, Bitterne's Santa hangs a sign on his tree reading 'SOS' ('short of sweets') and the neighbours hastily replenish his supplies.

'Perhaps you think we're all sick in the head, but we get a lot of enjoyment from seeing the children's faces when they see the real Father Christmas!'

SANTAS GROTTO

BUSY IN WORKSHOP!

Preparations start in July, but the actual hanging of lights takes only ten days with all able-bodied blingers needed to lend a helping hand. It takes up to four people to site the heavier items (such as a Santa's sleigh full of presents, full-size reindeer and man-size snowmen, to name but a few). All the houses are illuminated at 5.30 p.m. on the Saturday two weeks before Christmas. The neighbours hold a large, formal street turn-on, which brings out all the locals (and many visitors!) to line the pavements, listen to the pom-pom of a brass band and, of course, to watch the lights pop on. They used to hand out wine and mince pies, but in recent years it has become too large a gathering even for this generous bunch to accommodate. They have experienced much interest from both local and national TV stations and the lighting attraction has even been awarded a tourist board commendation for services to Southampton.

Mr N (dressed as Santa): What's your name, little boy?

Little boy: I told you last year!

Genre: Bungalow
Style: Fractal

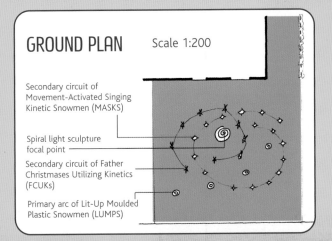

GROUND PLAN   Scale 1:200

Secondary circuit of Movement-Activated Singing Kinetic Snowmen (MASKS)

Spiral light sculpture focal point

Secondary circuit of Father Christmases Utilizing Kinetics (FCUKs)

Primary arc of Lit-Up Moulded Plastic Snowmen (LUMPS)

Charlton-on-Otmoor, Oxfordshire

Genre: Detached houses
Style: Mostly sophisticated

Three houses stand on a wide road in a pretty village setting, al doing their own bit of volunteer work to bring light to dark times The neighbourhood is dominated by the spectacular centra house, which exhibits a very practised display. Owners, the Ws regularly visit family in Georgia, USA, where the design of one' Christmas decorations is taken far more seriously than here Consequently, American Christmas Houses heavily influence these decorations. This may account for the careful design ethic: winter wonderland entirely made up of small fairy lights – no rope lights, no inflatables, no singing, and certainly no dancing. Colour are kept simple and create depth in the scene, trees are highlighted by glowing jewels and the centrepiece feature elegant reindeer sipping sparkling water from an icy pond.

The visit to America also influences the date the display is mounted – as late as the *second* week in December, as they do not return from their holiday until this time! 'We tend to ge

As if slightly bewildered, the neighbours immediately to the left have kept a sensibly low-key display in refined white. Conversely, those to the right seem to have reacted with overt patriotism and devoted themselves to determinedly British design ethics: make it big, make it random and slap a load of inflatables across your lawn. Marvellous!

Burnham-on-Sea, Somerset

It was a cold and lonely night when we pulled into the part of Burnham that people in the know had pointed us towards. Turning the corner, we were transported from the evening emptiness of a seaside town in winter, to a bustling wonderland of well-wrapped, neon pilgrims wandering around the close in awe.

A dozen large, detached houses faced towards the centre of the close where visitors congregated in a cloud of frosty-breathed gasps. The entire close had adopted a collective design strategy, eschewing all coloured lights and cartoon daftness and putting their full weight behind one approach: white only! Crisp, frost-white and pure as the driven snow.

Whether inspired by frosty nights, or restricted by lighting regulations protecting coastal shipping from being drawn on to the rocks, we couldn't tell. But silvery stars, icicles and ribbons of white light had turned the recognizable shapes of a house ... a porch ... a roof ... into ghostly landscapes of castles, cakes and, well, huge slabs of light.

The householders had developed styles varying from those of a palace ballroom to something more surreal. Yet they had all kept to the same pure colour principle. Silvery light beamed from all sides.

The effect was stunning.

Burnham-on-Sea, Somerset

Barton, Oxford

Genre: Terraced houses
Style: Fun fair

In this Noël Neighbourhood, there is a nice sense of fluidity to the lighting. Some collaboration amongst neighbours makes decorating easier – though there is an element of rivalry. Mr A is very jealous of Mr B's extremely long rope light, which he imported from America.

Mr A confessed that his electricity bill goes up fourfold over Christmas, despite the bulbs all being low voltage: 'It's because they're on all the time…It's flippin' expensive.'

How does he cope? Well, he has fewer lights this year. We're just sad we didn't visit last year.

THE SEASON *IV*
A NEW ROPE *LIGHT*

A long time ago in a *cul-de-sac*
far, far away from *the Galactic*
Empire...

It is the season of Scrooge and sneer, and a small band of
defiant revellers whose houses shine out from a previously hidden street,
have won their first victory against the evils of the shortest day.

Their secret weapon is this – they have created a space with enough power to
cheer an entire neighbourhood. As they race home each night to switch on,
pursued by the ever-earlier dusk, they are custodians to a secret knowledge
that could perhaps save their people from a sinister winter gloom...

The residents of this close in Buckinghamshire have a novel way of ensuring that they meet no opposition to their Christmas traditions from new neighbours. When Mr S moved here from Portugal – in the summer (!) of 1998 – he was given a welcome gift of a string of outdoor fairy lights from the other residents. Fortunately he was happy to grasp the challenge with both hands and the results speak for themselves (Mr S's house can be identified by its star-shaped bindi – see right). He is now an enthusiastic participant, who is developing a strong personal style.

The members of this community are happy to admit to being a little competitive and covetous of one another's new LED reindeer or blow-up snow shaker. Perhaps it is thanks to a thriving spirit of one-upmanship that the number of houses lighting up in this Noël Neighbourhood is still growing, year on year. It could be said that doing things out of season is also the local tradition, as last year the entire close threw an opening night *barbeque* in honour of the great switch-on.

Mr S's original string of lights sadly had a limited lifespan and has now been superceded by Movement-Activated Singing Kinetic Snowmen (or MASKS).

Kilgetty, Wales

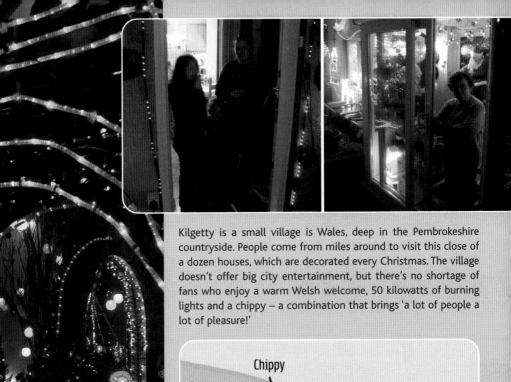

Kilgetty is a small village is Wales, deep in the Pembrokeshire countryside. People come from miles around to visit this close of a dozen houses, which are decorated every Christmas. The village doesn't offer big city entertainment, but there's no shortage of fans who enjoy a warm Welsh welcome, 50 kilowatts of burning lights and a chippy – a combination that brings 'a lot of people a lot of pleasure!'

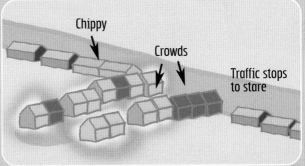

Chippy

Crowds

Traffic stops to stare

Decoration facts

Items and sets of lights used: 480
Units of electricity consumed per night: 200
Donations received up to Christmas Eve 2006 = £9,092.83
Donations received since 2000 = £54,000
Electricity bill over the Christmas period: £480
Fuse used: 100 amp (drawing about 90 amps)
Letters for Santa received each year: 2,000

Visitors will certainly be drawn to the corner house, which has been putting on impressive displays for a number of years. The decorations haven't gone unacknowledged by the wider community either. Creator Mr T has received the award for Best Dressed House in Wales for two years running (earning him a £100 voucher for Caerphilly Garden Centre!).

'It wasn't always quite so impressive,' Mr T says. He started small twenty-five years ago, but things somehow have got bigger every year. The annual preparation routine is no small undertaking. Each night for six weeks Mr T hangs his lights up round the kitchen from 4 a.m. onwards to test them, prior to installation. Out-of-season storage is also a major consideration for a display of such mammoth proportions and Mr T's 65 x 45 foot loft is choc-a-block with decorations year round

As if this were not enough, during the rest of the year, Mr T runs his own private museum with 6,500 items of Victoriana, domestic and transport memorabilia, prams, and First World War and Second World War artefacts.

Chapter 6

TRIPPING THE LIGHTS – FANTASTIC!

It is essential to have some dealings with the technical nitty-gritty when mounting your display.

If you start getting creative, making your own components or deviating from the off-the-peg options, it gets complicated. How on earth is it done? Collected here are some nerdy facts, entertaining construction details and comparative electrical consumption tables. It should be devastatingly dull, and yet...

Mr B sculpts his remarkable creations from recycled building materials, donated items and finds from car-boot sales. The first figure he created was in 1999 – a large snowman ingeniously knocked together using a water butt and a boat fender. Bolstered by the success of this experiment, more characters began to populate the forecourt.

The Snowies:

The beating heart of a Snowy is a windscreen-wiper motor, run through a transformer to bring down the voltage and its speed, and then adapted to create each band member's characteristic motion. Their frames consist of heat-moulded plastic strips, a breathable laminate-flooring underlay (to waterproof the internal electrics) and are covered with a geo-textile membrane outer. They have footballs for heads, their hats are upside-down buckets

How on earth?

The Snowies: Construction details

bubble wrap

geo-textile membrane outer

heat-moulded plastic strips covered in breathable laminate flooring underlay

football
pipe

windscreen-wiper motor

bucket
pipe lagging
hand-knitted accessories

SWAYING SNOWMAN   DRUMMING SNOWMAN   STRUMMING SNOWMAN with pipe-lagging brims, their noses are shaped pipes and they all wear scarves knitted by Mr B's mother-in-law. Their chubby snowman arms are shaped using layers of bubble wrap.

Santa's helpers:

The elves for Santa's workshop are based on the same mechanism; this time skip-destined, child-sized, shop mannequins donated by a children's clothing retailer are used. Again they have football heads, this time with masks to give them features. Apparently they looked 'a bit freaky' to begin with, but were less frightening once their beards had been stuck on.

The whole shebang is run off just two switched sockets (although to achieve this impressive economy, an area of garage approximately three square metres is covered by socket boards and cables). Mr and Mrs B experience an increase in their electricity bill of about £100 over the Christmas period (around seven weeks). They neatly compensate for this additional domestic expense by ensuring that the display wins the Detached House/Modest Mansion category of a local lighting competition each year, which covers the majority of the costs of mounting the display – 'mainly all the bloody transformers!'

'I take their hats off every year to blow their heads up a bit, then I silicon 'em back on!'

Equivalent daily energy consumption

× 90

× 4

× 1/2

× 900 slices

141

Mr C's dad is an electrician and has helped him to slowly upgrade the household electrics to make sure everything goes smoothly each year. In fact they've gone as far as installing an outside ring main under the decking so that they can plug in the glowing reindeer and inflatables nice and easily.

People are invited round to see the lights, expecting something small and go, "Oh my God!"

A Three-House Project

House A has bricked a scaffold pipe into their garage wall. This is in preparation for the high-drama moment when the neighbours manoeuvre their twenty-foot scaffold secret weapon, the 'T bar' into place. Read on to find out how they did it.

This wall of light transforms three neighbouring houses to closely resemble a cruise liner at full steam. The friends' work is very popular with the local community – they were inundated with complaints the only year they didn't put the lights up. In the past, the whole neighbourhood has come out to help celebrate the switch-on; one year they had television celebrity June Whitfield plugging them in. Local bands from the Army Cadets and Scout troops provided the entertainment and, to ensure it was safe and professional, they applied to the council to shut the road and provide public liability insurance.

C does all the roof work while his wife makes tea and sandwiches, leaving his neighbours, decorators A and B, to do the front.

How on earth did they join up the enormous gap between the two houses to achieve the desired ocean-liner effect? Serious engineering solutions were called for!

143

Stage one
Cast pipe into garage wall to construct bracket fixing (a)

Stage two
Construct scaffold T bar (b) using an Arc or MIG welder. Angle-grind all sharp edges smooth for safety.

Stage three
Using all available manpower, lower T bar (b) into bracket pipe (a).

This level of technical preparation is not always observed when dealing with those unruly cables. Many a decorator is happy to just wrap it in a bin bag.

4

Stage four

Illuminate!

Is your Christmas
House visible from space?

(To discover the answer, turn to p.190.)

Bore Your Friends!

Did You Know?

The father of Christmas tree lights as we know them today was Edward H. Johnson, an associate of the inventor Thomas Edison. Mr Johnson exhibited the first electrically lit Christmas tree in 1882 at his New York home, using patriotically coloured, hand-made bulbs.

Multicolour LED light sets sometimes require special wiring because the red and yellow coloured lights use less voltage than the blue-spectrum lights.

Icicle-shaped bulbs blow sooner than the more traditional variety.

The Queen has experimented with house-bling. On Christmas Eve 2003, a Union Jack was projected across the front of Buckingham palace, followed by a jolly geometric design of wrapping paper. But is it classy or kitsch, Ma'am?

A string of Christmas lights left on for ten hours a day over the Christmas period would produce enough carbon dioxide to fill fifty-two party balloons. ×52

A string of Christmas lights left on for ten hours a day over the Christmas period would use enough electricity to watch the Queen's speech over two hundred times. ×200

145

Chapter 7

TABLEAUX VIVANTS

As every student of French knows, *tableau vivant* means 'living picture'.

Tableaux vivants were a Victorian art form combining the traditions of theatre and painting, whereby a group of figures are appropriately (or inappropriately) costumed, posed and lit within a staged environment. Everybody stands very still, and the audience admires the composition or deciphers the story being told. Everybody claps. Sometimes photographs are taken to immortalize the image. Particularly popular in the nineteenth century were risqué *tableaux vivants* featuring nudes, and these were known as '*poses plastiques*'.

Some of our blingers have honoured this fine tradition by taking it in daring new directions, bringing a fresh interpretation to the term '*poses plastiques*'.

Warning: this chapter may feature nudes (if you count snowmen).

'Santa's Train's a-Comin"

An early forties Technicolor musical ensemble is recreated here, with all our
favourite characters helping Santa with the build-up to another busy Christmas.

Emotional finale: 11.5 out of 10

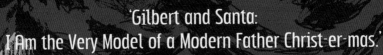

'Gilbert and Santa:
I Am the Very Model of a Modern Father Christ-er-mas.'

This Gilbert and Sullivan-inspired tableau stars not one, but two Santas in a mysterious case of mistaken identity. A battle of wits and a spot of vocal duelling ensue, sweetly accompanied by a chorus of angelic choirboys who beg for the truth to be revealed in time for them to hang their stockings.

Santa value: two for the price of one

'Huge Cat on Tiny Train'
'Ceci N'est Pas un Chat'

This shocking Surrealist juxtaposition plumbs the depths of our subconscious to raise disturbing dichotomies between the natural world and modern industrial age. Will the animal overpower the machine, or will the machine sweep the cat away in a blaze of light?

Surrealism out of ten: fish

'Frosty, the Snowmen'

This snapshot tableau from a lesser-known West-End musical is a heartfelt story of impoverished snowmen out in the harsh cold. Featured here is the big chorus number with all the cast members on the stage giving joyful thanks for newly donated hats and scarves.

Sob value: 7 out of 10

'It'll Be Lonely This Christmas'

Perhaps this tableau depicts the inherent loneliness of Santa's Christmas role. Father Christmas in his rocking chair has Homeric nobility as he contemplates the enormity of the task ahead. That he is dwarfed by this colossal throne with his job title blazing overhead could be said to be a metaphor for how any man would be overwhelmed by the responsibility of delivering all those presents in just one night.

Poor Santa. Cheer up, we'll give you 9 out of 10.

'Madonna with Child'

A classic Renaissance retelling of the greatest story ever told. Simple pictorial style. Note donkey resting at front.

Tradition value: 10 out of 10

Military Grotto 'Roswell'

This bizarre tableau explores themes of alienation in modern society. Note the discomfort of the isolated Victorian couple upon finding themselves confronted by harshly lit rocking horses and surrounded by a rusty iron fence. Tradition at odds with popular entertainment, perhaps? And we must ask ourselves why Santa and his little elves are enclosed in barbed wire. Just what is it that is so dangerous about Rudolph and the crew? Who is being protected – him or us?

Weirdometer grading: very

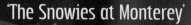

'The Snowies at Monterey'

A re-enactment of this near-mythical gig from the kaleidoscopic late 1960s.
Tragically the band's experimentation using lighter fuel for on-stage pyrotechnic
displays pushed their blazing careers into meltdown.

Rating in the Christmas charts: Number One!

'Untitled'

Here is a kitchen-sink drama illustrating the drudgery behind the glamour of the legend. No celebrity status for the ordinary workers shown here towards the end of a twenty-hour shift in mid-December. Clearly, they have not been given the chance to go home to wash or shave for several days and their clothes are hanging off their exhausted, shrunken frames. A moving portrait of the dedication and struggle behind a festival that brings joy to our children's faces every year.

Fair Trade rating: 2 out of 10.

'The Gingerbread House'

Paying homage to folkloric traditions of children's storytelling, this work exhibits a playful mix of scales. The cheerful scene constructs its own mythical society where giant animals (chipmunks) are neighbours with snowmen and Santas (or is it normal-sized chipmunks hanging out with miniature Santas?) The illuminated gingerbread house can also be read as an innovative response to the challenges of sustainable development, suggesting a form of 'high-calorie architecture' that offers shelter, sustenance and simultaneous treat value.

Sugar content: high

'House Party at Pingu's Place'

'Nuff said.

Pinguage: 85 per cent

Genre: Semi-detached house
Style: *Tableaux vivants* en masse

1. Dickensian prisoner dolls
2. Saintly magi visit the Christ child above, whilst heathen
   Santas hold a didgeridoo knees-up below
3. Victorian couple chat outside church

This house is one vast *tableau vivant*. Does it chart an epic tale of Christmas across the centuries, or were the participants caught in a north-west London flash flood and forced to take refuge on the garage? The creator mounts the entire display by himself and admits he finds it 'tedious and difficult' carting half-life-size nativity figurines on to his garage roof.

Participant score: cast of thousands

Chapter 8

COMMON PROBLEMS ENCOUNTERED

Sadly, not everything runs smoothly, even at Christmas. This chapter relates an assortment of challenges faced by some brave exhibitors, and how they coped when the Christmas spirit quietly absented itself from the proceedings.

1. ELECTRICAL DAMAGE

Long Hanborough

This experienced house-blinger has been decorating for thirteen years. Despite this impressive pedigree, she managed to blow her electrics last year. The damage seems to have had grave and far-reaching consequences – it has resulted in a *considerably* toned-down display this year.

2. INJURY

Powder foot

Most of the artists and craftsmen featured in these pages managed to hang their displays without too much damage to their persons. However, bringing to fruition the greatest of creative visions always entails a certain degree of risk. Two years ago, a Welsh gentleman we spoke to fell off his ladder on to his feet whilst taking down his display. Sadly this was not a graceful cat-like manoeuvre. His heel bone disintegrated into powder and as a result he was obliged to undergo a six-and-a-half hour operation to reconstruct his foot. 'I was packed in ice for two weeks!'

Happily he has recovered fully and is still able to out-class most blingers on the block.

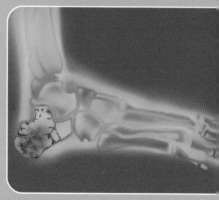

Theft is always a potential problem these days. At one house, the decorators now have a comprehensive CCTV system in place after thieves stole the donations box and broke some of the decorations. The owner multitasks by watching the CCTV monitor in tandem with his regular TV.

Juvenile thieves targeted another house on three separate occasions. The first time, the donations box was stolen from the front lawn leaving behind a Hansel and Gretel-style trail of copper coinage. The owner installed two CCTV cameras and movement-activated alarms. The following year, when kids attempted to rip the box from the bandstand, he was warned by the alarm, they were interrupted, and the whole incident was caught on film and passed to the police. Remarkably, another unsuccessful attempt was made the following year.

Perhaps a glimpse at the vastly impractical nature of the loot would help to discourage any future attempts. One homeowner told us he collects large sums for charity, but it is almost exclusively in two-pence pieces. This means he banks around thirty bags of change two or three times a week and has had to open a special bank account to cope. Not an easy haul to shift, even in the sweetie shop.

It's not always about money. At 2 a.m. on Christmas morning following TV coverage of a third house, a large plastic snowman and a musical snowman were 'literally ripped' from the front lawn and stolen. The owners were scandalized and very upset. They continue to mount a display showing multiple snowmen, but these days 'in very different positions'.

4. NEIGHBOUR PROBLEMS

MILD

The creator of this exhibit has five grandchildren he could use as an excuse, but blatantly admits that he really puts the lights up for himself. A neighbour complained about light pollution, so he put more up: 'I'm a bit rebellious like that.'

MODERATE

Here is a display that has been heavily influenced by problems with neighbours. Bizarrely, the majority of the close is lit up, but this house has always been the outstanding beacon. The toned-down decorations this year are due to many complaints about the numbers of visiting fans turning their cars around at the end of the close. Previously, the exhibitor collected substantial donations for a local school but his neighbours' unofficial use of parking cones resulted in fewer cars stopping and he was forced to withdraw this practice. In the past he became Santa in residence, greeting visitors and dispensing gifts accompanied by Molly the dog dressed up as a reindeer. However, he no longer has the heart to do so. He felt obliged to lower the bulb count, so now the sides of the house are not decorated. Instead, he projects a large Santa on to the house of his supportive next-door neighbour.

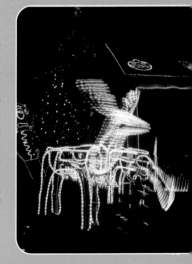

'I'd like to get him to put more lights up there on my house next year.' – The next-door neighbour.

EXTREME: The ASBO House

A detached house on a private road in an upmarket area (exact location not given for legal reasons).

An ASBO has been taken out against these Christmas lights which have been declared a menace to the local neighbourhood. In past years, a spectacular light display drew in visitors from across the country. In 2006, however, the local district council won an injunction against the creator, citing neighbour complaints, traffic snarl-ups, increased levels of crime, and a £7,400 bill for policing the immediate neighbourhood. The dizzying light display and many enthusiastic visitors have also been blamed for events as diverse as the break-up of a marriage and the severe disorientation of an individual viewing the display.

After a £36,000 court case and sixteen hours in prison, this blinger has major restrictions on what he is entitled to display. The height of inflatables is limited to seven feet, any noise frowned upon and overall light levels restricted. (Apparently, a £15.99 snowman from Argos has caused more problems since the injunction.) Any violations of these restrictions, or conversations with journalists will result in the condemned man going to prison for seven days. As a result, donations to charity Daisy's Dream have dropped from £54,000 over the past five years to a mere £250 this year.

'This is all I'm allowed this year, and my Winnie the Pooh is still three inches too tall for the terms of the court injunction...and I've got fourteen light bulbs too many.'

Many decoration enthusiasts lamented the passing of one particular illuminated house near Brighton. During its heyday, this house brought the road to a standstill with coach parties, fireworks, Salvation Army bands, the Brighton and Hove Albion football team, and mince pies

and sherry. It was also reputed to have snow machines on the roof that showered visitors when they helped themselves to humbugs from a jar near the door.

The owner admits that he had a concealed camera so he could time it just right for the first snow machine: 'Then they looked up and we got them with second one! They got covered!'

One visitor from a local nursing home had tears in her eyes, explaining that she 'hadn't smiled in ten years since [her] hubby died', and it does sound like it was a massive glowing party.

However, the hundreds of pounds-worth of sweets, lights, Bristol Cream and the visitors ringing the doorbell till midnight was not something that could go on for ever. Eventually this decorator decided to retire gracefully and pass the torch on to other decorators around the country.

PIMP MY SLEIGH

THAT WARM FEELING INSIDE

House decorators like to share their love of light, so of course most of the effort goes on the outside of the house. It is a selfless act – once at home, the artist can no longer see his creation. However, there is a dedicated band of house decorators whose enthusiasm cannot be limited to the exterior. They continue the extreme glow action into the centre of their world. Step into the light!

A glowing palace stands on the hill, part-semi, part-space ship. It seems this house has recently landed after its journey from the North Pole (via Mars). Giant beings Homer Simpson and Tigger have stepped outside to sample the atmosphere. Captain Mr C and his dog First Mate Rex are responsible for this outlandish outpost of otherworldliness.

Inside the vessel, the fantasy theme continues. The lounge has been transformed into a wintry snowscape, enhanced by the existing white decor of the walls and sofa.

The icicles, snowmen, a frosty tree and a snow-covered forest obviously go down well with our wolfy first mate. Perhaps it makes him think of bounding across the open icy plains of Scandinavia, back on Earth, hunting for rabbits and snow squirrels.

Mr K and Mrs G live opposite each other in the same street. Although they are quick to play down any rivalry, it is difficult to not see these houses as trumpeting a decorative challenge to one another.

Of the two, Mrs G is the newcomer, having only put up lights for the last seven years.

She was born and bred in this part of Coventry and was so inspired by her friends Mr and Mrs K, that she thought she should have a go too. She says her friends have been very encouraging with ideas and technical tips.

Festive decorations have been common inside our houses for centuries. Dressing your house in lights to brighten up the winter streets for whoever passes by takes things a little further. But to bring that same intensity back in through the front door, to completely transform your lounge with decorations that can be seen half a mile away, extends the whole concept into uncharted territory.

Both these householders have done just that.

On the other side of the road Mr and Mrs K are the more experienced decorators. It takes them six to eight weeks to put up their display, which at last count consisted of over 20,000 lights. After twenty-five years of displays, the fame of their decorations has spread and visitors drop by from far afield. The house is featured on television and in newspapers every year, and there is a steady stream of tinsel-pilgrims from across the country and far beyond. The year before last, they were pleased to welcome a visitor from Nairobi, who was drawn by their star.

However, these two households have taken things that little bit further still...

Step inside either of these houses and you enter an Aladdin's cave of shimmering warmth. Although she may still have a way to go to rival her neighbour on the outside, Mrs G has really done her finest work inside the house. The cloud of lights and tinsel isn't the end of it – a life-size, singing Santa adds that extra-special touch.

Meanwhile, over the road, a constant stream of visitors is invited inside to wonder at how the decorations continue over the threshold. Tinsel reflects hundreds of twinkling fairy lights dancing up the walls and across a star-spangled ceiling. Decorations hang down not unlike a reflective foil jungle and light comes from all sides.

There's a power to these decorations that is somehow difficult to describe. With a room that has as many bulbs as this, you would think everything would be clearer inside. However, the effect is spellbinding. The eye doesn't quite know what to focus upon. These house decorators have dispelled all the shadows from their houses and what the visitor experiences is a hypnotic mist of light that casts a mesmerizing effect upon entry. Wherever you look it is bright and soft, a mirage of festive imagination made real.

Chapter 11

CHRISTMAS EVERY DAY

There are a chosen few who retain an essence of the season all year round.

Chelsea, London

Genre: Georgian terrace
Style: Otherworldly

Even in summer, even in *daylight*, this house would cause any passer-by to stop and gawp. Every inch of the façade is hidden behind tangled vines and leafy shadows, and the door of the house peeps out as though set back in a fairytale forest. The mere metre of space between the house wall and the pavement is built up into a rocky mountain kingdom populated with fantasy creatures – white horses pull a golden coach, unicorns graze cheek-by-jowl with giant kingfishers, and there are noble stags, huge fluffy bunnies and something that looks like a skunk (but could be a badger).

This remarkable and original house has a permanent display of live and artificial plants and the fantasy inhabitants create a living grotto every day. The upper floors of the façade house a moderate-size aviary. Over the Christmas season, there is even more. More fairies than usual twinkle from amongst the greenery and more reindeer winter in the rockery.

The spirit of Christmas can always be found here, whether it is December or July. The sense of transporting us gently to an enhanced reality where magic is entirely possible can be gathered just by looking. The owner of this house declared the purpose of the decorations was to educate and amuse visiting children, and we can only assume they would have been delighted by this proof of truth in fairytales.

Unfortunately this fantasy vision can no longer be seen and has vanished back into the dreamland from whence it came. But those of us who had the privilege of seeing it remember the creator for the pleasure it brought us.

The Real Mrs Christmas

Worthing, West Sussex

Mrs Christmas says that her son doesn't really like the decorations. 'But it's only once a year,' she tells him.

Most of the lights come from a garden centre on the A259 and even with forty boxes in the attic she still picks up a few more every year. The overall design may vary, but there's always a tree at the centre of the display.

'I do love my tree,' she explains. She used to have one inside, but birds landed in it (she's got a lot of birds) and the dog knocked it over with his tail (he's an energetic dog), so now she puts a tree up outside, to share it with everyone.

What with power-surge plugs in the shed and a time-switch in the porch, there's a lot of wiring outside. Luckily, Mrs Christmas has a great solution to keep things dry. She puts the main socket in a rabbit hutch, wrapped in polythene.

The Self-styled Mr Christmas

Melksham, Wiltshire

Residents of the Wiltshire town of Melksham will need no further introduction to the festive antics of Mr P, a self-styled Mr Christmas who makes regular appearances in the pages of the local paper and on regional radio. For the forty-seven-year-old electrician, it just isn't enough to have one festive high point in his year. He has decided to make every one of his days Christmas Day. His exhausting routine involves a full Christmas dinner, crackers, presents and Queen's speech 365 days a year. A little arithmetic suggests that since 1993 he's feasted on over a metric tonne of turkey, 120,000 sprouts (can that be possible?) and 900 litres of sherry. He claims his mince pie count is up to 250 a week (that would be 10 tonnes in total).

Does he do something special to mark 25 December?

'No, for me it's just a normal day.'

Of course this extreme performance attracts a lot of attention, and he has hosted local radio shows from his front room, as well as having entertained the big names of entertainment (including the great Sir Cliff Richard). A couple of years ago he released a Christmas single to promote his rather quirky philosophy of celebration, a 'Radio 2-style' musical offering put together with the help of Peter Andre's producer entitled 'It's Christmas Every Day'. Unfortunately, our research efforts have been unable to locate any evidence of how well it did in the charts.

So *This* Is Christmas?

Here's to all those bizarrely out of
season items that slink in with the
other coloured lights!

What is it about Christmas that so many people love?

It is very easy to lose a feel for the season and declare it solely for small children, especially if family get-togethers are difficult. But Christmas can offer a doorway through which we can step into a fantasy for a few weeks, and access that glowing, hazy place between childhood and reality.

Maybe Christmas reminds us of a child's anticipation of wild dreams fulfilled by a bulging stocking. Maybe the visual warmth of twinkling lights seeps in to make our spirits glow. Either way, the atmosphere is special. Reality is softened and faded by luminous colours and light-hearted, familiar characters and sounds. Perhaps the more one decorates, the more solid the dream becomes.

See your Christmas house from space!

To work out how much bling you will require,
follow this simple formula:

1 Watt = 11 lumens emitted in a sphere

100 W = 1,100 lumens emitted in a sphere

Surface area of a sphere = $4\pi r^2$ (π = 3.1415926535...)

∴ Output for 100 W bulb = Total output ÷ bulbs surface area = $1,100/4\pi r^2$

A steradian is a cone of light, spreading out from the bulb

1 lumen per steradian = 1 candela

1 candela can be seen for 30 miles with the naked eye

Our bulb has $4\pi r^2 \times 1^2 = 12.6$ steradians

So, for an output of 1 candela in each steradian you need 12.6 lumens

One 100 W bulb has 1,100 lumens which is the same as 1,100 ÷ 12.6 candelas

As a result a single 100 W electric light bulb can be seen from (1,100 ÷ 12.6) x 30 = 2,619 miles away

Now this sounds quite a lot, but of course this is if your lights are displayed in a darkened vacuum, and not everyone's will be, so these other factors need to be taken into account:

Background light pollution – Get away from everyone else and find a nice dark desert to install your lighting display (preferably at high altitude). Even then make sure it's a new moon.

Atmospheric interference – varies according to weather conditions, but we'd guess a 100 W light can be seen from about 5 miles in average conditions (so divide your total by 524).

Final formula:

(Christmas House wattage x 26.2) ÷ 524

No one can actually agree how far space is, so we'll say 100 miles (which is the lowest orbit for a satellite, and you'd probably need one of those to see your lights anyway). If your answer is more than 100, you can see your house from space, well done!

The authors would like to thank the following for their help/enthusiasm/patience/all three: Sarah Such (it really, really wouldn't have happened without you!); Sal, Ivy and Mira; the house-spotters: Rufus, Sid and Steve; Poppy for her Coventry expertise and Richard and Alison for their Yorkshire help; Professor Anton Lazarov of the Burgas Free University, Bulgaria and Dr Duncan for simplifying the formula; Dee, Lou and Will.

Thanks to the Energy Saving Trust for some entertaining energy consumption comparisons.

Huge thanks for all the warm welcomes we've received on cold winter evenings at the dazzling houses in this book, and special thanks to Dave and Marian of Southampton, Graham (Smith) of Wigmore and Phillip and Susan of Denton for all their additional research, photos and DVDs showing the history of their neighbourhoods.

We've always loved ogling lit-up houses – keep it up, please.

Gutted that we missed your spectacular display? We are too, so send us your pics at www.christmashouses.info, which also has more Christmas House related links than you can shake a snowman at.